Contents

Abstract

Screaming Trees: The Nigerian Deforestation Crisis

Nigeria has the highest rate of deforestation in the world and the Government of Nigeria (GON) has failed to implement an effective response to this worsening crisis. Deforestation degrades land quality and agricultural output, leading to forced migration and increased competition for scarce resources. This erodes government legitimacy and, ultimately, leads to conflict that results in more deforestation. The GON and international community must act to break this cycle and reverse deforestation to prevent widening conflict and irreversible environmental degradation. This paper begins with an analysis of the linkage between deforestation and conflict, followed by an overview of the underlying environmental and socio-economic conditions that enable deforestation. Next is an explanation of the relationship between deforestation, climate change, and Nigerian quality of life. Building on this foundation, the analysis shifts to Nigerian efforts to address the deforestation crisis, with a specific focus on corruption and poor governance. Finally, this paper concludes with recommendations for Nigeria and the international community to reverse deforestation through a holistic, integrated campaign that is focused on economics, education, governance, and innovation.

INTRODUCTION

"I am mindful that I represent the shared aspiration of all our people to forge a united Nigeria: a land of justice, opportunity, and plenty."

President Goodluck Jonathan, 2011

When the USAFRICOM Commander, General Carter Ham, testified before the Senate Armed Services Committee on the status of Nigeria on February 29, 2012, he focused on a range of security matters, including recent activity among violent extremist elements such as Boko Haram, piracy concerns in the Gulf of Guinea, and regional security cooperation efforts.[1] What he did not address, however, was the Nigerian deforestation crisis, which represents an existential threat to the entire West African ecosystem.

Nigeria has the highest rate of deforestation in the world and lost half its forest cover over the past two decades.[2] The resulting deterioration of land quality has contributed to the encroachment of the Sahara desert into northern Nigeria, where 1,350 square miles of land is lost to the desert each year.[3] Deforestation has degraded Nigerian quality of life through its adverse impact on the agricultural sector and this crisis will continue to intensify competition for scarce resources among the Nigerian people. The Government of Nigeria (GON), in cooperation with the international community, must reverse deforestation trends in order to preclude an environmental crisis that could heighten the risk of conflict.

Deforestation is the catalyst for a vicious cycle in Nigeria: deforestation leads to environmental degradation and forced migration, resulting in competition for shrinking resources and increased tension among the rural poor. This sets the conditions for conflict and instability that leads to even more deforestation and environmental degradation. This cycle is accelerating, and the environmental consequences could be devastating. The UN estimates, for example, that 2/3 of arable land in Africa could be lost to desertification by

2025.[4] To date, however, the U.S. has committed relatively few resources in preventive foreign aid to assist Nigeria with this impending disaster.[5] This is a Nigerian problem that will require a Nigerian solution, but the GON will need substantial help to solve this crisis.

This paper will first analyze the linkage between resource shortages and conflict, and then provide an assessment of the underlying environmental and socioeconomic conditions in Nigeria that contribute to deforestation. Next is an examination of the relationship between deforestation, climate change, and Nigerian quality of life, followed by an evaluation of Nigerian efforts to address the deforestation crisis. Finally, this paper concludes with recommendations for the GON and international community to reverse deforestation.

Environmental Degradation and Regional Conflict

Over the past six decades, at least 40% of all intrastate conflicts involved disputes over natural resources.[6] Direct linkage between environmental factors and conflict is difficult to prove, as there are always additional underlying causes, but in Nigeria there is a compelling argument that deforestation is a primary catalyst for instability.[7] Land and water shortages resulting from deforestation lead to forced migration and the loss of agricultural jobs. This unemployment reduces quality of life, with a disproportionate effect among the bulging youth population under the age of 24, whose jobless rate approaches four times that of Nigerians who are older than 45.[8] The inadequate government response to the deforestation crisis has eroded government legitimacy, especially when considered in the context of wealth disparity. The Nigerian GDP grew at a rate of 6.8% from 2005-2011 and the oil industry continues to thrive, yet over 60% of the population lives on less than $1 per day.[9] The cumulative effect of these conditions is an unstable environment that is a potential powder keg for conflict, as depicted in Figure 1 below. [10]

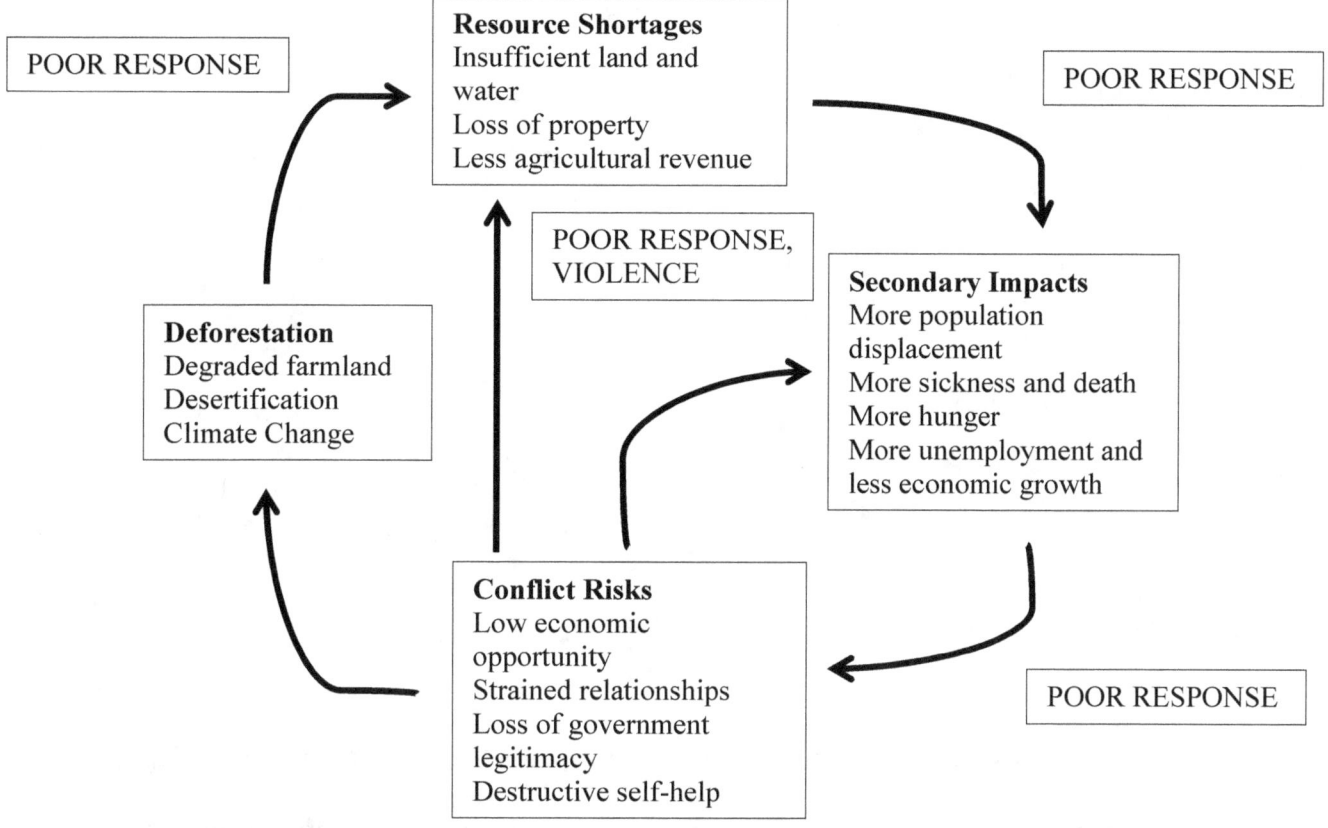

Figure 1. Deforestation and Conflict in Nigeria (adapted from Sayne, Aaron, *Climate Change Adaptation and Conflict in Nigeria,* p. 3, [Washington, D.C.: United States Institute of Peace, 2011])

Resource-driven conflict is erupting across Nigeria. The shortage of arable land in

northern states leads farmers and herders to migrate to unfamiliar lands, driving conflict in

areas where Muslim nomadic cattle-herders wander onto land occupied by Christian

farmers.[11] Displaced environmental refugees in the Lake Chad region are also experiencing

widespread conflict due to the lack of adequate water and food that has resulted from the

extended drought in northeast Nigeria over the past decade.[12] The potential consequences of

resource-driven conflict are evident in Darfur, where land shortages and famine contributed

to open warfare that killed over 300,000 people and displaced at least two million others.[13]

Communal violence across Nigeria over the past 10 years is already responsible for the deaths of 10,000 citizens, and much of this conflict resulted from resource disputes.[14]

In failing states, environmental degradation leads to a loss of government legitimacy: the government loses popular support when the people perceive a governmental failure to recognize an environmental crisis or to offer a credible solution.[15] The symptoms of this condition are overtly present in Nigeria, where poor governance and high unemployment have led to popular resentment. The Nigerian people have low confidence in the GON's ability to provide equitable distribution of oil revenue and they resent the consistent failure of the GON to implement effective environmental solutions.[16] Thus the vicious deforestation cycle continues in Nigeria, and the people are suffering the consequences.[17]

The Deteriorating Nigerian Environment

Having addressed the relationship between resource shortages and conflict, we now turn to a specific examination of the environmental and socio-economic context in which deforestation occurs, then review its causes and effects. The Nigerian environment is a complex adaptive system in which the relevant actors and their effects on each other continuously change and defy predictability. The key actors include the GON, the international community, the corporate sector, the people, and the environment itself. Any successful intervention must consider the unintended consequences for each of these actors, and should account for their continued interaction, rather than focusing on the individual actors themselves.

With that in mind, the most constant features of the Nigerian socio-economic fabric are paradox and instability. Nigerian GDP is 41st highest in the world, largely due to oil wealth, but Nigeria ranks 161st in the world in earnings per capita.[18] Population growth is also inversely proportional to job growth; while Nigerian population (the highest in Africa) continues to grow at 3% annually, unemployment is approaching 25%.[19] The youth represent the sector most affected by high unemployment, as 60% of the Nigerian population and 75% of its unemployed citizens are under the age of 30.[20] This unemployed youth bulge represents enormous potential for growth, but also a great source of volatility.

The agricultural sector is the traditional bedrock of the Nigerian economy; 70% of Nigerian workers derive their livelihood from agriculture, yet it is this sector that is most at risk to the effects of environmental degradation.[21] Deforestation is leading to worsening drought conditions, water shortages, rising temperatures, and severe reductions in land that is suitable for planting. As a result of these trends, projections indicate that the Nigerian agricultural sector could suffer annual losses of up to 62% going forward.[22] While the oil sector continues to grow, it is not reducing unemployment rates or increasing wealth for most Nigerians.[23] Thus, due to the loss of traditional agricultural employment and the imbalanced oil economy, jobless Nigerian youth are forced to migrate into an overcrowded urban environment that features limited job opportunity and high susceptibility to unrest.

This destructive cycle begins with deforestation, broadly defined as the removal of trees as a result of human or natural causes (natural disasters, fires, etc.).[24] Since 1990, Nigeria lost half of its overall forest cover and over 75% of its primary forests, which are the ecologically rich forests that are untouched by human development or natural disaster.[25] Due to a lack of verifiable data, the specific causes of Nigerian deforestation are difficult to pinpoint.[26] Most studies show, however, that the principal causes of deforestation are subsistence farming and the harvesting of trees for fuel wood that is used for cooking and heating. Fuel wood is the main source of domestic energy, which is ironic in a country that is one of the world's largest oil producers.[27] According to the Food and Agriculture Organisation (FAO), "In Africa 90% of wood removals are used for energy and a reversal of the situation would depend on structural shifts in economies to reduce direct and indirect dependence on land."[28]

Other causes of deforestation include illegal logging, residential construction resulting from population growth, and widespread displacement of herders who remove trees to create pastures to replace those areas lost to the expanding desert.[29] Deforestation rates are accelerating and reforestation efforts have been negligible; GON replanting schemes, for example, have only resulted in a 4% replacement rate of new seedlings compared with disappearing forestland.[30]

Closely related to deforestation is desertification, which is the long-term transformation of forestland into desert, a phenomenon that occurs in the arid regions in northern Nigeria, when land is neglected after trees are felled without replacement.[31] Based on current environmental trends in Nigeria, 2/3 of 11 northern states could turn to desert or semi-desert during the 21st century.[32] Similar conditions are already evident on the northeast border of Nigeria, where Lake Chad has shrunk by 90%, and the Sahara is swallowing 351,000 hectares of potential farmland each year.[33] When previously arable land is lost to desertification, it is almost impossible to resurrect that land for future agricultural use.[34]

The Nigerian culture offers some explanations for irresponsible deforestation practices. For example, the poor educational system in Nigeria results in a low level of awareness among the Nigerian public regarding the insidious effects of deforestation, and the people are unaware of any government regulation that prohibits the free removal of trees. The GON does not require states to incorporate environmental education into school curricula and, in any case, Nigerian children have limited opportunity to attend public school.[35]

The GON also has a poorly developed regulatory structure for forest management, and only 6% of forests have regulatory protection.[36] Nigerian citizens regard forestland as a "Common Pool Resource" which should be free from government intervention.[37] When managing "common pool resources," individuals tend to be motivated by their own gain at the expense of the well-being of the environment and are less likely to manage natural resources in a responsible fashion.[38] Rural farmers who are forced to migrate because of land degradation are less apt to properly care for unfamiliar land than their own ancestral property. The Niger Delta has suffered greatly from this phenomenon, due to the increased migration of Northern farmers to the Delta region, where they have ravaged the mangrove forests.[39]

The Nigerian economy, then, is characterized by imbalance and wealth disparity; most Nigerian citizens do not benefit from oil wealth, but instead suffer high unemployment as a result of the deteriorating agricultural sector. The agricultural sector is at risk because of environmental degradation that begins with deforestation, and these environmental challenges are only worsening as a result of inadequate education, destructive Nigerian cultural norms, and the absence of regulatory oversight.

Deforestation, Climate Change, and Quality of Life

> *"The life of man, solitary, poor, nasty, brutish, and short."*
> *Thomas Hobbes*

Deforestation has degraded Nigerian quality of life by accelerating the rate of climate change and urbanization, resulting in high jobless rates and reduced standards of living. Deforestation without replenishment begins a destructive pattern of soil degradation that results in the loss of farmland and biodiversity and the release of a carbon footprint that is second only to the burning of fossil fuels as a contributor to climate change.[40] The effects of climate change are evident across Nigeria, where there are fuel wood shortages, disappearing hunting grounds, rainfall decreases of 3-4% per decade, rising sea levels and temperatures, water shortages, and desertification.[41]

Deforestation in Nigeria accounts for 87% of total carbon emissions, and Nigerian temperatures are increasing at a rate that is 50% greater than the mean rise in global temperature.[42] The adverse impact of deforestation on climate change has placed a large sector of the Nigerian population at risk and, conditions are worsening. Only 20-25% of rural communities and 45-50% of urban communities have access to safe drinking water and Nigeria currently ranks 156 of 187 on the Human Development Index that measures overall

8

standards of living such as life expectancy, literacy, and education.[43] Climate change most directly impacts the poor, who are dependent on environment-related sources of income, and evidence shows that climate change could force an additional 220 million people into poverty across the globe by the end of the 21st century.[44] Flooding related to sea-level rise will also displace millions of Nigerians, especially those residing in low-lying coastal areas.[45]

There is a bright side to climate change, however. Given the worldwide attention that has been granted this issue, there is an opportunity for Nigeria to receive financial benefit from participating in efforts to reduce global warming. The "Reduced Emissions from Deforestation and Forest Degradation (REDD)" program, for example, is a vehicle for compensation in return for reduced rates of deforestation.[46] Some states have begun to cooperate with this program and the GON has instituted modest plans for agricultural adaptation, but results thus far have been limited and localized.[47]

The unfortunate reality is that worsening deforestation has continued to degrade Nigerian quality of life and the GON response has been inadequate. Nigerians who can no longer find work in the agricultural sector are flocking to coastal cities such as Lagos, where low-lying urban areas are particularly vulnerable to rising seas, and resentment is rising among slum-dwellers who live in the shadow of skyscrapers.[48] The GON has the responsibility to reverse this vicious cycle, but governmental actions to date have been ineffective, as the next section will show.[49]

Assessing the Government Response

"Corruption is the greatest single bane of our society today. Nigeria must change its ways in order to ensure progress, justice, harmony and unity and above all to rekindle confidence amongst our people."

President Obasanjo, 1999

Nigeria has no shortage of organizations responsible for environmental oversight, and the GON has tried to implement programs and initiatives to address environmental degradation, yet there is scant evidence of progress. Nigeria sits at 14 of 20 on Foreign Policy's "failed states index," indicative of the corruption and inept bureaucracy that have hindered environmental restoration. [50] The Nigerian people view the government as an economic "black box" into which revenues and aid are deposited and out of which little progress emerges. Poor ecological governance and corruption is the backdrop for the destructive cycle of Nigerian deforestation; environmental degradation continues unabated, poverty is getting worse, and the people are losing faith in the government.

The Nigerian Federal Ministry of the Environment (FME) is the highest national agency responsible for environmental oversight and encompasses the Federal Environmental Protection Agency (FEPA); each Nigerian state possesses its own environmental protection agency, and most actions are delegated to the state level.[51] There are also a host of NGOs and IGOs engaged in deforestation efforts; most specifically, the Nigerian Environmental Study/Action Team (NESAT).[52] The FME recently published a "National Adaptation Strategy and Plan of Action on Climate Change for Nigeria (NASPA-CCN)," a promising treatise on Nigerian environmental challenges that offers a well-rounded environmental strategy, but the GON has failed to implement this strategy or establish valid assessment mechanisms for progress.[53] Unfortunately these environmental agencies are also infected with corruption; popular perception is that leaders of these agencies regard incoming environmental funding as "pocket money or election funding."[54]

No environmental solution will be effective without the implementation of significant anti-corruption measures. Corruption takes many forms in Nigeria: criminal misuse of funds for personal gain, malfeasance, and in some cases incompetence.[55] The result is the same: money provided to the government is not used to fix the problem, resulting in a loss of faith in the government and worsening environmental conditions.[56] Corruption has depleted as much as 40% of Nigeria's annual income and GON officials have testified that they cannot properly account for as much as 65% of annual funding.[57]

Poor governance and administration share the blame for GON's failure to manage the deforestation crisis. Nigeria is one of the world's largest oil producers, with GDP growth at 6.8%, yet it remains among the world's poorest countries.[58] The GON has failed to address wealth disparity or to implement equitable revenue-sharing measures; the current level of revenue sharing in Nigeria is 13%, which is insufficient to meet the needs of the poor majority.[59] The GON has also mismanaged its human capital and failed to leverage the rich capability that exists within the Nigerian population to fulfill important requirements for labor, innovation, and entrepreneurship. Efforts to fight deforestation require manpower-intensive projects such as replanting campaigns, and manpower is one resource that Nigeria has in great quantity.

When faced with environmental challenges, the adaptive potential of human beings is limitless; in Niger, for example, farmers adapted to extreme drought by developing a successful plan to regenerate tree stumps that reinvigorated planting in an arid region.[60] There have been similar instances of reforestation in other under-developed nations such as Vietnam, where the Vietnamese Red Cross worked with citizens in a successful reforestation campaign to replant 2,000 hectares of mangrove trees.[61]

Because the GON has not provided credible regulation or leadership on environmental matters, however, Nigerian citizens adopt "destructive self-help" practices, in which they assume a quasi-governmental role of environmental management in the absence of a legitimate regime of oversight.[62] This situation is analogous to the U.S. National park service relinquishing oversight and regulation of national assets such as the Grand Canyon; it is not difficult to envision the destructive practices that would result if the American people and public corporations were permitted to have free reign over those resource-rich lands.

Additionally, the GON has failed to leverage its vast oil resources to fund alternative energy solutions for rural citizens who are depleting the forests for fuel wood. Furthermore, the GON has neglected to reinvest oil revenues into infrastructure improvements such as roads and power-lines that would enable distribution to rural areas. As a result, poor Nigerian farmers cut down trees for fuel wood because they lack an alternate energy source. The Nigerian gas-flaring crisis is another example of a missed opportunity to translate Nigerian resources into an alternate energy solution; Nigeria ranks second in the world in gas flaring, which has resulted in the wasteful release of over 13 million barrels of oil into the environment.[63] This is a missed opportunity for the GON to capture these wasted resources to help fund or provide alternate energy to the rural poor.

The GON is also failing to optimize the potential for oil revenues to support reinvestment in technology and education. Such reinvestment could contribute to environmental solutions, especially if undertaken in cooperation with international and corporate entities. Unfortunately, the GON has not taken steps in this direction; instead, tree-planting programs have been negligible and a failing educational system has not produced the necessary expertise in fields such as forestry, biology, and engineering that could enable innovative solutions that are so important in this environment.[64]

The absence of proper governmental oversight and regulation in Nigeria, in combination with widespread corruption, ensures the deforestation problem will only get worse without significant reform. The Nigerian people do not exercise responsible environmental practices because they do not see the point. The GON has not educated the people regarding the importance of sound environmental practices, nor has it taken the necessary steps to reverse environmental degradation. As a result of these failures, the Nigerian government has lost legitimacy in the eyes of the masses and exacerbated a volatile situation that could lead to conflict.

Is this really such a big deal?

Some would argue that there are indications the GON has implemented an effective program to manage deforestation. GON agencies and NGOs are containing environmental degradation in some communities, the environment is resilient, and the booming Nigerian oil economy could offset the negative effects of deforestation. Programs such as the Great Green Wall Sahara Initiative (GWSI), for example, offer promising resources that regional actors are employing to address deforestation, desertification, and climate change. The GWSI has been trumpeted by President Goodluck Jonathan as an initiative that will restore farmland and arrest desertification in the Sahel region on the Northern Nigerian border.[65] While the GON recognizes its problems associated with wealth disparity and economic imbalance, the world's top investment banks believe it is feasible that Nigeria can become one of the world's top 20 global economies by 2025.[66]

Some communities in northern Nigeria have shown environmental resilience and increases in per-capita agricultural production during the latter portion of the 20th century, and there are similar examples of successful environmental adaptation in resource-limited countries such as Ethiopia.[67] While there is some internal unrest and environmental degradation is cause for concern, the government has built organizations such as the FEPA who are monitoring the situation, providing adequate oversight, and may possess the capacity to mitigate the deforestation crisis before it leads to widespread conflict. The GON has also overseen the creation of the NASPA-CCN that encompasses a National Forest Policy and is a comprehensive strategy to address deforestation.[68]

While it may be true that the GON has acknowledged the deforestation crisis and taken some measures to implement a solution, it is delusional to believe that this problem is under control. A critical review of the NASPA-CCN, for example, reveals no clear mechanism for funding allocation or any realistic assessment schedule.[69] It is encouraging that the GON has developed organizations to manage the deforestation challenge, but actions to date have been ineffective and data reveals that environmental degradation is accelerating. Deforestation continues unabated, quality of life is falling, and without dramatic near-term action, the deforestation crisis could lead to irreversible environmental damage and conflict. Going forward, it is urgent that the GON and the international community pursue immediate solutions to stabilize the environment in order to prevent such consequences.

CONCLUSION

"If you want to go quickly, go alone. But if you want to go far, go together."

African Proverb

Nigerian forests are disappearing at the highest rate in the world, resulting in the rapid decay of the Nigerian environment, regional instability, and increased potential for conflict. Hampered by corruption and ineffective bureaucracy, the GON has failed to resolve this crisis. Meanwhile, the Nigerian people are faced with a disappearing agricultural livelihood, reduced quality of life, and eroding faith in the Nigerian government. In order to reverse these conditions and prevent conflict, the GON and international community must recognize the scale of the problem and develop collaborative solutions in the near term. The right way forward is a global and integrated approach that leverages the potential of Nigerian human capital, with a clear focus on the objective of improving Nigerian quality of life and achieving stability through the reversal of deforestation trends. The GON will need strong leadership and transparency in a campaign that is focused on actions in the areas of Economics, Education, Governance, and Innovation in order to turn the tide of deforestation. Organized and led under such an effective plan, the Nigerian people have the ability and resources to meet this challenge, prevent environmental catastrophe, and preserve regional stability.

RECOMMENDATIONS

"The pessimist sees difficulty in every opportunity. The optimist sees the opportunity in every difficulty."

Sir Winston Churchill

A solution to Nigerian deforestation will require an integrated campaign across multiple Lines of Effort (LOE): Economics, Education, Governance, and Innovation. In order to succeed, this campaign will require widespread participation from Nigerian states and communities, and a high degree of international support.[70] Most importantly, the Nigerian population must assume a primary role in supporting any government action and the GON should work to provide the people with the expertise and responsibility to solve this crisis. While the GON must lead any solution, the international community cannot stand idly by and "hope" for progress. The GON must be incentivized to pursue progressive policies, and must understand that it will be held accountable for failure.

15

Economics: Economic prescriptions should be focused on the young and vibrant Nigerian human capital that is the engine of Nigerian economic growth, offers the labor needed for the work force, and the ingenuity needed to solve this problem. A free market approach that using programs such as REDD will generate revenue to fund improvements in infrastructure and other implementation costs.[71] The GON must institute economic reform to reduce wealth disparity and achieve more equitable distribution of resources via state and local governments. The GON should also encourage the agricultural growth by providing free access to credit for Nigerian farmers, and the Nigerian population should be energized in a nationwide tree-planting campaign to reverse the tide of deforestation.[72] U.S. and international aid to Nigeria must increase, but should be contingent on government reform that entails dramatic anti-corruption measures. The GON should also partner with corporate entities to ensure that companies who benefit from Nigerian natural resources are doing their part to fund and assist with reforestation.

Education: Because of low public awareness regarding the environmental impact of deforestation, Nigeria should incorporate a system of "ecological governance" that focuses on education and re-formulation of societal norms in favor of forest preservation.[73] This methodology should employ strategic communications, information tools, and school curricula as vehicles to educate the Nigerian people about negative impact of deforestation and to empower citizens to participate in the solution. Educational themes should highlight the process and benefits of crop rotation, reforestation, and efficient grazing methods. NGOs and IGOs will play important roles in educating the population and the GON should co-opt such organizations for this purpose.[74]

Governance: There is no singular organization or agency that is effectively managing the deforestation crisis, but this does not imply that government expansion is needed to solve this problem. Organizationally, the problem may be the existence of ***too many*** Nigerian governmental organizations with overlapping responsibilities and authorities. The GON will have to find its own solution, but a rational approach would be to streamline existing governmental organizations into a lean, efficient environmental protection agency with clear regulatory authority and minimal redundancies from the national to the community level. Environmental regulation reform must take place in order to arrest the rate of deforestation. Corporate entities should be held accountable for penalties associated with violating forestry regulations, and the GON must increase the percentage of protected forestland. Improved transparency is essential for the GON to offset the domestic and global perception of rampant Nigerian corruption and to clear the way for increased foreign investment.

Innovation: Nigeria has a massive labor force, but lacks expertise in the fields of engineering, technology, and forestry that will be necessary to solve this complex problem. The GON must leverage its organic human capital in pursuit of innovative environmental solutions, in collaboration with the global community of NGOs, IGOs, and corporate entities that will play an important role in providing expertise and technological capability. The GON should pursue alternate energy sources such as electricity and solar power for the rural poor to reduce the use of trees for fuel wood, and must work to find solutions for the gas-flaring crisis. The GON also must work with IGOs to obtain accurate assessment of deforestation rates in order to baseline future efforts and support qualification for climate-change related funding streams. While external organizations will be key enablers of

17

innovation, the ultimate goal should be to grow the expertise and capability within Nigeria to set the conditions for long term success and improved Nigerian quality of life.

NOTES

[1] Commitee on Armed Services, United States House of Representatives, *Statement of General Carter Ham, U.S. Army, Commander United States Africa Command*, February 29, 2012.

[2] B.A. Usman and L.L. Adefalu, *Nigerian Forestry, Wildlife and Protected Areas: Status Report*, Tropical Conservancy Journal: Biodiversity II (3&4), 2010, 49. Nigeria's overall forest coverage has shrunk from 20% to 10%.

[3] N.A. Onyekuru and Rob Marchant, *Nigeria's Response to the Impacts of Climate Change: Developing Resilient and Ethical Adaptation Options*, Agriculture and Environmental Ethics, 18 August 2011, 587.

[4] Olanrewaju Fagbohun, *Environmental Degradation and Nigeria's National Security: Making Connections*, Lagos: Nigerian Institute of Advanced Legal Studies, 2011, 368.

[5] Aaron Sayne, *Climate Change Adaptation and Conflict in Nigeria*, United States Institute of Peace, Special Report 274, June 2011, 13. While the EU and Canada have contributed aid to address climate change, U.S. contributions have been negligible. USAID/Nigeria Strategy 2010 – 2013 does not allocate funding for anti-deforestation initiatives.

[6] Silja Halle, *From Conflict to Peacebuilding: The Role of Natural Resources and the Environment*, United Nations Environment Programme, 2009, 5.

[7] Ibid., 8. "Environmental factors are rarely, if ever, the sole cause of violent conflict. Ethnicity, adverse economic conditions, etc. …are also correlated. However, it is clear that the exploitation of natural resources and related environmental stresses can become significant drivers of violence."

[8] U.S. Department of State, Nigeria Economic Fact Sheet, http://nigeria.usembassy.gov/nigeriafactsheet html/ (accessed 30 Sep 12). Unemployment rate among Nigerians between15-24 is 41.6%, while it is 11.5% among 45-59 year olds.

[9] Ibid.

[10] Ibid., 5

[11] Sayne, 5.

[12] Fagbohun, 367.

[13] Halle, 9.

[14] Sayne, 5.

[15] Fagbohun, 369. "Environmental degradation can cause people to lose faith in and become discontent with their leadership's ability to govern them, promote development, provide basic goods and services, and create a prosperous national economy."

[16] Sayne, 7. There is a trend of recent violence against state entities in Nigeria by groups who claim governmental failure as justification. This has especially been the case in the Niger Delta, an area of minimal governance and widespread environmental degradation.

[17] Sayne, 13. Conflict in the Niger Delta has added an additional cost of $3- $16 per barrel of oil.

[18] Ibid., 11.

[19] Ibid., 11. At current growth rates, Nigerian population could be 741 million by 2100.

[20] Ibid., 6.

[21] Onyekuru, 586. Nigeria is projected to lose between 42% and 60% of agricultural GDP as a result of climate change.

[22] Ibid., 586.

[23] William Ehwarieme, *Corruption and Enviromental Degradation in Nigeria and its Niger Delta*, Journal of Sustainable Development in Africa, Vol 13, No. 5, (2011), 39.

[24] Dennis Anderson, *The Economics of Afforestation*, (Baltimore: The Johns Hopkins University Press, 1987), 10. Effects of tree loss include erosion, water table and soil degradation, and the terminal phase of desertification.

[25] Grace Azibuike, "Nigeria: Deforestation – Looming Self-Inflicted Disaster", 11 September 2011. http://allafrica.com/stories/201109110144.html, 2. (accessed 15 Sep 12)

[26] Nathaniel Olugbade Adeoye, *Assessment of Deforestation, Biodiversity Loss and the Associated Factors: Case Study of Ijesa-Ekiti Region of Southwestern Nigeria*, GeoJournal, 6 Jan 2010, 231.

[27] Azubuike, 1.

[28] Ibid., 2.

[29] Ehwarieme, 41. Thisday news reported that "more than 37 percent of the country's forest reserves were lost between 1990 and 2005 as a result of illegal and uncontrolled logging, incessant bush burning, fuel wood gathering, and clearing of forests for other land uses."

[30] Adeoye, 231. Reforestation only accounts for 25,000 ha each year, while 600,000 ha of forest is destroyed.

[31] Edith C. Pat-Mbano, *Climate Change Reduction: A Mirage in Nigeria,* Canadian Research and Development Center of Sciences and Cultures, 9 Mar 2012, 15.

[32] Sayne, 4.

[33] I. J. Ekpoh, *The Effects of Recent Climatic Variations on Water Yield in the Sokoto Region of Northern Nigeria*, International Journal of Business and Social Science, Vol. 2 No. 7, April 2011, 251.

[34] Anderson, 39.

[35] Usman, 44.

[36] Illegal Logging.info, Tropical Forest Trust, last modified October 29, 2012, http://www.illegal-logging.info/approach.php?a_id=102. (accessed 30 Oct 12)

[37] Murtala Mohammed, "Combating Desertification in Northern Nigeria: The Need for Paradigm Shift," http://murtalaadogi.wordpress.com/2012/01/22/combating-desertification-in-northern-nigeria-the-need-for-paradigm-shift/, 22 Jan 2012.

[38] Ibid., 2.

[39] Ewharieme, 38. "Migrants are not interested in resource conservation and environmental protection since their main motive is profit maximization."

[40] Lenny Bernstein, et. al., *Climate Change 2007 Synthesis Report*, Intergovernmental Panel on Climate Change Nov 2007, 36.

[41] Onyekuru, 587.

[42] Azubuike.

[43] Ehwarieme, 41. Components of the Human Development Index include life expectancy, literacy, education, and standards of living.

[44] Pradosh K. Nath, *A Critical Review of Impact of and Adaptation to Climate Change in Developed and Developing Economies*, Journal of Sustainable Environmental Development, 25 Jun 2010, 142. The 2006 "Stern Report" forecasts that by 2100, an additional 145-220 million additional people will exist below the $2-a-day poverty line as a result of climate change.

[45] Bernstein, 50. Nigerian population, currently 160m, is forecasted to be over 700m by 2100. Lagos will be extremely vulnerable to the effects of sea rise under those conditions.

[46] Christian Nelleman, *Green Carbon, Black Trade: Illegal Logging, Tax Fraud and Laundering in the World's Tropical Rainforests*, United Nations Environment Programme, 2012, 5.

[47] Government of Nigeria, *Federal Ministry of Environment, National Adaptation Strategy and Plan of Action on Climate Change for Nigeria (NASPA-CCN)*, Dec 2011, 35. GON has also developed the "Building Nigeria's Response to Climate Change (BNRCC) plan, which addresses climate change considerations in detail. Cross River State has been active in seeking participation in REDD initiatives.

[48] Sayne, 5. Over nine million homes may be threatened by rising seas resulting from climate change by 2050.

[49] V.T. Jike, *Environmental Degradation, Social Disequalibrium, and the Dilemma of Sustainable Development in the Niger-Delta of Nigeria*, Journal of Black Studies, Vol. 34 No. 5, May 2004, 691. Oil companies have also neglected to assist with afforestation efforts; nor has the GON held oil companies accountable.

[50] U.S. Africa Command testimony, 3. "Fragile states lack the capacity or political will to effectively address demographic, political, social, and economic challenges, including population growth, urbanization, income equality … increasing demand for internal resources."

[51] NASPA-CCN.

[52] Onyekuru, 590.

[53] NASPA-CCN.

[54] Ehwarieme, 44. "You can hardly get to any state and see any ecological project of note that the governor has done with the money collected."

[55] Ehwarieme, 35. Out-of-law corruption is "action outside the confines of law … illegal and punishable." In-law corruption is more insidious and less overt, described as "serving state interests at the expense of the public through inefficiencies." Corruption is often subtle and results more from inefficiency and redundancy than criminal activity; for example, there are two Nigerian government organizations that exist to address natural disasters, the Federal Disaster Relief and Protection Agency and the Ecological Fund.

[56] Ibid, 37.

[57] Ibid., 42.

[58] U.S. Department of State, Nigeria Economic Fact Sheet, (accessed 1 Nov 12).

[59] Kelly Campbell, "*Bringing Peace to the Niger Delta*," United States Institute of Peace, June 2008, 93. Ehwarieme also notes that while revenue to the Niger Delta region increased by 63% between 2006 and 2008, this did not translate into any measurable improvement in Nigerian environmental conditions.

[60] Nath, 154.

[61] Ibid., 156.

[62] Sayne, 7. "Those who see the state as weak or self-serving may choose to fashion their own responses to climate change. State failures could encourage more crime."

[63] Pat-Mbano, 15.

[64] Usman, 51.

[65] Vanguard Nigeria News, *FG launches GWSI project to curb desert encroachment*, last modified September 11, 2011, http://www.vanguardngr.com/2012/09/fg-launches-gwsi-project-to-curb-desert-encroachment.

[66] Catherine S. M. Duggan, *Nigeria: Opportunity in Crisis?*, Harvard Business School Publishing, Aug 2009, 1.

[67] Nath, 150.

[68] NASPA-CCN, 1.

[69] Ibid., 38.

[70] Ehwarieme, 45.

[71] Sayne, 13. GON should also develop a National Adaptation Program of Action (NAPA), which supports the UN Framework for Climate Change (UNFCC) and is a requirement to access UN funding for climate change.

[72] Gibbs, 1.

[73] Ehwarieme, 45. M. M'Gonigle defined "ecological governance" as a restructuring of societal systems in a manner that offers alternatives to extractive, linear, and unsustainable systems that level forests.

[74] Azubuike.

BIBLIOGRAPHY

Adeoye, Nathaniel Olugbade. *Assessment of Deforestation, Biodiversity Loss and the Associated Factors*: *Case Study of Ijesa-Ekiti Region of Southwestern Nigeria*. GeoJournal, 6 January 2010.

Anderson, Dennis. *The Economics of Afforestation*. (Baltimore, MD: The Johns Hopkins University Press, 1987).

Awe, F. et al. *Impact of Deforestation on the Economic Activities of People in Okun Area of Kogi State, Nigeria*. Agricultural Economics, 2012.

Azubuike, Grace. "Nigeria: Deforestation – Looming Self-Inflicted Disaster." 11 September 2011. http://allafrica.com/stories/201109110144.html. (accessed 15 Sep 12)

Bailey, Shawn T. *Climate Change, Instability and a Full Spectrum Approach to Conflict Prevention in Africa*. Newport, RI: US Naval War College, 23 Oct 2009.

Bello, O.S. *The effects of over cultivation on some soil properties, nutrients response and yields of major crops grown on acid sand soils of Calabar South-Southern part of Nigeria*. Brisbane, AU: 19th World Congress of Soil Science, 2010.

Bernstein, Lenny et al. *Climate Change 2007 Synthesis Report, Intergovernmental Panel on Climate Change*. November 2007.

Campbell, Kelly. "*Bringing Peace to the Niger Delta*." United States Institute of Peace, June 2008.

Committee on Armed Services, United States House of Representatives. *Statement of General Carter Ham, U.S. Army, Commander United States Africa Command*, 29 February 2012.

Committee on Foreign Relations, United States Senate. *International Deforestation and Climate Change*. Hearing Before US Senate Subcommittee on International Development and Foreign Assistance, Economic Affairs, and International Environmental Protection, 22 April 2008.

Duggan, Catherine S. M. *Nigeria: Opportunity in Crisis?* Harvard Business School Publishing, August 2009.

Ehwarieme, William. *Corruption and Enviromental Degradation in Nigeria and its Niger Delta*. Journal of Sustainable Development in Africa (Vol 13, No. 5), 2011.

Ekpoh, I. J. *The Effects of Recent Climatic Variations on Water Yield in the Sokoto Region of Northern Nigeria*. International Journal of Business and Social Science, Vol. 2 No. 7, April, 2011.

Fagbohun, Olanrewaju. *Environmental Degradation and Nigeria's National Security: Making Connections*. (Lagos: Nigerian Institute of Advanced Legal Studies, 2011).

Halle, Silja. *From Conflict to Peacebuilding: The Role of Natural Resources and the Environment*. United Nations Environment Programme, 2009.

Government of Nigeria. *Federal Ministry of Environment, National Adaptation Strategy and Plan of Action on Climate Change for Nigeria (NASPA-CCN)*, Dec 2011.

Gibbs, Holly K. *Monitoring and Estimating Tropical Forest Carbon Stocks: Making REDD a Reality*. IOP Science Journal, Aug 2007.

Illegal Logging.info. Tropical Forest Trust, last modified October 29, 2012. http://www.illegal-logging.info/approach.php?a_id=102. (accessed 29 Oct 12)

Jike, V.T. *Environmental Degradation, Social Disequalibrium, and the Dilemma of Sustainable Development in the Niger-Delta of Nigeria*. Journal of Black Studies, Vol. 34, No. 5, May 2004.

Maier, Karl. *This House Has Fallen*. (New York: Public Affairs, 2000).

Mohammed, Murtala Adogi. *Combating Desertification in Northern Nigeria*. Last modified September 11, 2011. http://murtalaadogi.wordpress.com/2012/01/22/combating-desertification-in-northern-nigeria-the-need-for-paradigm-shift/. (accessed 15 Sep 12)

Nath, Pradosh K. *A Critical Review of Impact of and Adaptation to Climate Change in Developed and Developing Economies*. Journal of Sustainable Environmental Development, 25 Jun 2010.

Nelleman, Christian. *Green Carbon, Black Trade: Illegal Logging, Tax Fraud and Laundering in the World's Tropical Rainforests*. United Nations Environment Programme, 2012.

Onyekuru, N.A. and Rob Marchant. *Nigeria's Response to the Impacts of Climate Change: Developing Resilient and Ethical Adaptation Options*. Agriculture and Environmental Ethics, 18 Aug 2011.

Pat-Mbano, Edith C. *Climate Change Reduction: A Mirage in Nigeria*. Canadian Research and Development Center of Sciences and Cultures, 9 Mar 2012.

Sayne, Aaron. *Climate Change Adaptation and Conflict in Nigeria*. United States Institute of Peace, Special Report 274, Jun 2011.

Sharma, Narendra. *Managing the World's Forests: Looking for Balance Between Conservation and Development*. (Dubuque, Iowa: Kendall/Hunt Publishing Co., 1992).

Usman, B. A. and Adefalu, L. L. *Nigerian Forestry, Wildlife and Protected Areas: Status Report*. Tropical Conservancy Journal: Biodiversity II (3&4), 2010.

U.S. Department of State. *Nigeria Economic Fact Sheet*. http://nigeria.usembassy.gov/nigeriafactsheet.html/ (accessed 30 Sep 12).

Vanguard Nigeria News. *FG launches GWSI project to curb desert encroachmen.*, Last modified September 11, 2011. http://www.vanguardngr.com/2012/09/fg-launches-gwsi-project-to-curb-desert-encroachment/. (accessed 30 Sep 12)